Harmony in Green: Exploring Earth's Ecosystems and Our Environmental Responsibility

In the lush tapestry of our planet's landscapes, a symphony of life unfolds in harmonious shades of green. "Harmony in Green: Exploring Earth's Ecosystems and Our Environmental Responsibility" invites you on a profound journey through the interconnected web of life that encompasses our world. From the towering forests to the delicate grasslands, from the depths of the oceans to the canopy of the rainforests, this book is a celebration of the intricate relationships that shape Earth's ecosystems.

In an era where our actions have far-reaching consequences, the need to understand and safeguard our environment has never been more pressing. This book seeks to bridge the gap between ecological wonder and environmental responsibility. Through its pages, you will embark on an expedition that traverses continents, dives into oceans, and soars among the clouds, all while delving into the heart of our interconnected natural world.

As we explore the diverse realms of Earth's ecosystems, we will encounter the myriad species that call them home. From the smallest microorganisms to the majestic apex predators, each organism plays a vital role in maintaining the balance of their respective ecosystems. Alongside these vivid depictions of biodiversity, we will delve into the intricate relationships that weave life together – predator and prey, pollinator and plant, symbiotic partners that underscore the delicate equilibrium of our planet's ecosystems.

Yet, "Harmony in Green" goes beyond the awe-inspiring beauty of nature. It serves as a call to action, a reminder of our collective responsibility to preserve and protect the world we share. The chapters will explore the myriad ways in which human activities have impacted these ecosystems, from deforestation and pollution to climate change and habitat destruction. Through an understanding of these challenges, we are empowered to make informed choices that promote sustainability and regeneration.

But this book is not just a plea for change; it's an anthem of hope. "Harmony in Green" showcases stories of successful conservation efforts, innovative solutions, and the resilience of nature itself. It highlights how communities, scientists, and individuals are collaborating to restore ecosystems, preserve biodiversity, and foster a renewed sense of balance between humanity and the natural world.

In a world that often feels overwhelmed by global challenges, "Harmony in Green" reminds us that we are all stewards of this planet. It emphasizes that every action, no matter how small, contributes to the greater harmony of Earth's ecosystems. This book is an invitation to embark on a transformative journey – one that enriches our understanding, ignites our passion for preservation, and reinforces our commitment to leaving a thriving legacy for generations to come.

As we turn each page, may we find inspiration in the intricate interplay of life, embrace the responsibility of safeguarding our environment, and ultimately, discover our place within the symphony of "Harmony in Green."

Introduction: The Lush Symphony of Earth's Ecosystems

- Earth's ecosystems and the interconnectedness of life.
- Importance of understanding and protecting the environment for future generations.

Chapter 1: Environmental Science: Unveiling Earth's Complex Systems

- Environmental science and its interdisciplinary nature.
- Scientific methods used to study and understand Earth's ecosystems.

Chapter 2: Environmentalism: A Call to Protect Our Planet

- History and evolution of the environmental movement.
- Profiles of key environmentalists and their contributions to global awareness and action.

Chapter 3: Conservation: Preserving Biodiversity and Ecosystems

- Conservation and its importance in maintaining balance in ecosystems.
- Case studies of successful conservation efforts and the positive impacts they've had.

Chapter 4: Ecology: The Web of Life and Interdependence

- Ecological principles and the relationships between organisms and their environments.
- Food chains, food webs, and the role of keystone species.

Chapter 5: Recycling: From Waste to Resource

- Significance of recycling in reducing waste and conserving resources.
- Innovative recycling programs and the benefits of adopting sustainable practices.

Chapter 6: Water Supply: Navigating the Challenges of a Precious Resource

- Importance of clean water sources for both human and ecological health.

- Water scarcity, pollution, and the global efforts to ensure safe water access.

Chapter 7: Weather: Unpredictable Patterns and Climate Change

- Earth's climate system, atmospheric phenomena, and weather patterns.
- Impact of climate change on weather events and the planet as a whole.

Chapter 8: Desertification: Battling the Spread of Arid Lands

- Causes and consequences of desertification.
- Initiatives to combat desertification and restore degraded lands.

Chapter 9: Natural Disasters: Nature's Unpredictable Fury

- Occurrence and causes of natural disasters like earthquakes, hurricanes, and wildfires.
- Disaster preparedness, mitigation strategies, and community resilience.

Conclusion: A Harmonious Future: Nurturing Earth's Ecosystems

- Importance of environmental responsibility and individual actions.
- Embrace a harmonious relationship with nature and become stewards of the environment.

Epilogue: Our Ongoing Journey: Shaping the Future of Earth's Ecosystems

- Ongoing efforts to protect and restore Earth's ecosystems.
- Potential for positive change and collective action.

Earth's ecosystems and the interconnectedness of life

Earth's ecosystems are intricate and interconnected networks of living organisms and their physical environments. These ecosystems encompass a wide range of environments, from lush rainforests and expansive oceans to arid deserts and towering mountain ranges. What makes ecosystems truly remarkable is the delicate web of relationships that tie every living and non-living element together.

Interdependence of Life: In Earth's ecosystems, every living organism, no matter how small or seemingly insignificant, plays a role in maintaining the balance of the system. Plants, animals, fungi, and microorganisms interact with one another in complex ways. For example, plants use sunlight to perform photosynthesis, producing oxygen and serving as the foundation of the food chain. Herbivores consume plants, while carnivores feed on herbivores. This interdependence creates a chain of relationships that sustains life.

Biodiversity's Crucial Role: Biodiversity, or the variety of life forms within an ecosystem, is a cornerstone of its stability. A diverse range of species helps maintain ecosystem resilience in the face of environmental changes. Each species has adapted to its niche over countless generations, resulting in a tapestry of life that contributes to nutrient cycling, pollination, disease control, and other vital processes.

Energy Flow and Nutrient Cycling: Energy flows through ecosystems in a hierarchical structure, with producers (plants) capturing sunlight and converting it into chemical energy through photosynthesis. This energy is then transferred to consumers (animals) as they feed on plants or other animals. Decomposers, such as bacteria and fungi, break down dead organisms, returning nutrients to the soil for plants to use once again. This cycling of energy and nutrients forms the basis for all life within the ecosystem.

Habitat and Niche: Each organism within an ecosystem occupies a specific habitat and niche. A habitat is the physical environment where an organism lives, while a niche refers to the role that organism plays within the ecosystem. Niche includes factors like an organism's diet, behavior, and interactions with other species. No two species occupy the same niche without some form of competition or adaptation.

Keystone Species and Trophic Cascades: Certain species within an ecosystem have a disproportionately large impact on the balance of the ecosystem. These are known as keystone species. For example, the presence or absence of a top predator can influence the abundance of prey species, leading to cascading effects throughout the ecosystem. When keystone species are removed, it can disrupt the delicate equilibrium and result in unforeseen consequences.

Human Impact and Conservation: Human activities have profound effects on ecosystems. Deforestation, pollution, overfishing, and climate change alter the delicate balance of ecosystems, threatening species and habitats. Conservation efforts aim to protect biodiversity, restore degraded areas, and preserve the interconnected relationships that sustain life on Earth.

Understanding the interconnectedness of life within Earth's ecosystems is essential for both ecological health and human well-being. By recognizing our role within these intricate systems and promoting sustainable practices, we can work toward maintaining the harmony and balance that make Earth's ecosystems so vibrant and awe-inspiring.

Importance of understanding and protecting the environment for future generations

Understanding and protecting the environment is not only crucial for the present but also holds immense significance for the well-being and sustenance of future generations. Here's why:

1. Preserving Biodiversity: Biodiversity is the foundation of Earth's ecosystems, providing a wide range of resources, including food, medicine, and raw materials. A loss of biodiversity can disrupt ecosystems, leading to potential food shortages, decreased resilience to disease, and reduced availability of essential resources for future generations.

2. Climate Stabilization: Maintaining a healthy environment is vital for regulating the Earth's climate. By curbing activities that contribute to climate change, such as greenhouse gas emissions, future generations can avoid the severe consequences of rising temperatures, sea level rise, extreme weather events, and disruptions to agriculture and water resources.

3. Sustainable Resources: Many natural resources, such as clean water, fertile soil, and forests, are essential for human survival and well-being. Overexploitation of these resources can lead to scarcity and environmental degradation. Responsible management and conservation today ensure that these resources will be available for future generations to meet their needs.

4. Ecosystem Services: Ecosystems provide a wide array of services that benefit humans, such as pollination, water purification, and soil fertility. These services are critical for agriculture, health, and overall quality of life. Protecting ecosystems ensures that these services continue to benefit future generations.

5. Ethical Responsibility: As caretakers of the planet, there's an ethical responsibility to leave a habitable and thriving world for those who come after us. We have the power to make choices that positively impact the future, and failing to do so could lead to irreversible consequences that burden future generations.

6. Health and Well-Being: A healthy environment directly contributes to human health and well-being. Clean air, water, and natural spaces are essential for physical and mental health. By protecting these elements, we ensure that future generations can enjoy a high quality of life.

7. Cultural and Natural Heritage: Cultural and natural heritage are intertwined with the environment. Protecting ecosystems and landscapes means preserving the rich cultural history and natural beauty that can inspire and enrich the lives of future generations.

8. Technological and Scientific Innovation: Many technological innovations are inspired by nature. Studying ecosystems and natural processes can lead to discoveries that help address future challenges, such as renewable energy solutions and sustainable agricultural practices.

9. Avoiding Environmental Debt: Current environmental degradation can create a burden of debt that future generations will have to address. By taking proactive steps to reduce pollution, conserve resources, and adopt sustainable practices, we avoid passing on an environmental debt to our descendants.

10. Legacy and Responsibility: The legacy we leave for future generations is not just about material possessions; it's about the health and sustainability of the planet. By prioritizing environmental protection and sustainable living, we demonstrate a responsible approach to caring for Earth and its inhabitants.

In summary, understanding and protecting the environment are critical for ensuring that future generations inherit a world that is both habitable and abundant in resources. By acting today to address environmental challenges, we empower future generations to thrive in a world that respects and safeguards the interconnected web of life on Earth.

Environmental science and its interdisciplinary nature

Environmental science is a multidisciplinary field that seeks to understand the interactions between the natural world and human activities. It draws from various scientific disciplines to address complex environmental issues, explore the impacts of human actions on the environment, and develop solutions for a sustainable future. The interdisciplinary nature of environmental science allows for a comprehensive and holistic approach to studying and addressing environmental challenges. Here's how different disciplines contribute to the field:

1. **Biology:** Biology plays a crucial role in understanding the diversity of life on Earth, including ecosystems, species interactions, and adaptations. Ecological studies examine how organisms interact with each other and their environment, shedding light on topics such as population dynamics, food webs, and biodiversity conservation.

2. **Chemistry:** Chemistry is essential for analyzing pollutants, understanding chemical reactions in the environment, and assessing the quality of air, water, and soil. Environmental chemists study how pollutants move through different compartments of the environment, their transformation processes, and their potential impacts on ecosystems and human health.

3. **Geology and Earth Sciences:** Geology contributes by examining the Earth's physical structure, processes, and history. Geologists study rock formations, landforms, soil composition, and

geological hazards like earthquakes and volcanoes. Their research helps predict and mitigate natural disasters and provides insights into the long-term behavior of Earth's systems.

4. Physics: Physics is applied in environmental science to study energy dynamics, radiation, and atmospheric processes. It's critical for understanding climate change, energy sources, and the behavior of waves and particles in the environment.

5. Ecology: Ecology explores the interactions between organisms and their environment, ranging from individual species to entire ecosystems. This discipline provides insights into the complex relationships that shape the distribution and abundance of species, nutrient cycles, and the stability of ecosystems.

6. Atmospheric Science and Meteorology: These disciplines focus on the Earth's atmosphere, including weather patterns, climate, and atmospheric composition. Environmental scientists in this field study air quality, atmospheric pollutants, climate change, and weather-related events.

7. Sociology and Anthropology: Sociology and anthropology contribute by examining human behavior, culture, and social structures that influence how societies interact with the environment. This includes understanding the factors that drive environmental attitudes, behaviors, and policy decisions.

8. Economics: Environmental economics evaluates the costs and benefits of human activities on the environment. It assesses the economic value of ecosystem services, such as clean air and water, and explores strategies for sustainable resource management.

9. Political Science and Policy Studies: These disciplines delve into the governance and policy frameworks related to environmental issues. Environmental policy experts analyze the development, implementation, and effectiveness of policies aimed at addressing environmental challenges.

10. Engineering and Technology: Engineers and technologists design solutions to environmental problems, such as developing renewable energy sources, designing efficient waste management systems, and creating sustainable infrastructure.

The interdisciplinary approach of environmental science allows researchers and practitioners to tackle complex issues from multiple angles. It recognizes that environmental challenges are intertwined with social, economic, and scientific factors, and that holistic solutions require collaboration across various fields of expertise. By integrating insights from these disciplines, environmental science contributes to a deeper understanding of the environment and guides efforts toward sustainable practices and policies.

Scientific methods used to study and understand Earth's ecosystems

Studying and understanding Earth's ecosystems involves a systematic and rigorous scientific approach. Scientists use various methods and techniques to gather data, analyze patterns, and draw conclusions about the complex interactions within ecosystems. Here are some key scientific methods used in the study of Earth's ecosystems:

1. Field Observations: Field observations involve direct data collection in natural environments. Scientists observe and record interactions between organisms, behaviors, population sizes, and physical factors like temperature, humidity, and soil composition. Long-term ecological research sites provide valuable data on ecosystem changes over time.

2. Surveys and Sampling: Sampling methods involve collecting data from a subset of the ecosystem to infer characteristics of the entire ecosystem. Random sampling, systematic sampling, and stratified sampling are commonly used to collect representative data on species distribution, biodiversity, and other ecosystem parameters.

3. Remote Sensing: Remote sensing uses satellite imagery, aerial photography, and other technologies to collect data from a distance. It's valuable for monitoring large areas and tracking changes over time, such as deforestation, urban expansion, and changes in land cover.

4. GIS (Geographic Information Systems): GIS combines spatial data (maps) with attribute data (information about the features on the map) to analyze and visualize patterns in the environment. It's used to create maps, model spatial relationships, and analyze ecosystem changes.

5. Modeling: Ecosystem modeling involves creating computer simulations to understand how different factors interact within an ecosystem. Models can predict the effects of environmental changes, population dynamics, and the impacts of management strategies.

6. Data Analysis: Statistical analysis helps scientists identify patterns, correlations, and trends in collected data. Techniques such as regression analysis, ANOVA (analysis of variance), and multivariate analysis are used to draw meaningful conclusions from complex datasets.

7. Laboratory Studies: Laboratory studies involve controlled experiments conducted in controlled environments. Scientists can manipulate variables and observe how ecosystems respond to changes in conditions, such as temperature, pH, or nutrient availability.

8. Isotope Analysis: Isotope analysis helps trace the movement of nutrients and elements within ecosystems. Stable isotopes can reveal information about food sources, nutrient cycling, and the movement of species.

9. DNA Analysis: DNA analysis is used to study biodiversity by identifying species through their genetic material. DNA barcoding and metagenomics provide insights into species composition and genetic diversity within ecosystems.

10. Citizen Science: Citizen science involves involving the public in data collection efforts. Apps and online platforms allow individuals to contribute observations of plants, animals, and environmental changes, which can provide valuable data for researchers.

11. Long-Term Monitoring: Long-term ecological monitoring involves continuous data collection over extended periods. This approach helps scientists identify trends and changes that occur over years or decades.

These methods, often used in combination, contribute to a comprehensive understanding of Earth's ecosystems. They enable scientists to reveal the intricate relationships between species, the impact of human activities, and the complex dynamics that shape the health and resilience of ecosystems.

History and evolution of the environmental movement

The history and evolution of the environmental movement span centuries and have been shaped by various cultural, social, and scientific developments. The movement has grown from isolated efforts to a global force advocating for environmental protection, conservation, and sustainability. Here's an overview of its key milestones:

1. Roots in Conservation and Preservation (19th Century):

- Early conservationists like George Perkins Marsh (Man and Nature, 1864) and John Muir (founder of the Sierra Club) highlighted the importance of preserving natural landscapes and resources.

2. Emergence of National Parks and Wildlife Sanctuaries (Late 19th - Early 20th Century):

- The establishment of Yellowstone National Park in 1872 marked the world's first national park, inspiring the conservation of public lands for recreational and natural purposes.
- Teddy Roosevelt's presidency (1901-1909) expanded protected areas and focused on conservation policies.

3. Environmental Awareness and Regulation (Mid 20th Century):

- Rachel Carson's "Silent Spring" (1962) raised awareness about the dangers of pesticides and their impact on ecosystems and human health.

- The first Earth Day was held on April 22, 1970, catalyzing widespread public engagement in environmental issues.
- The United States Environmental Protection Agency (EPA) was established in 1970 to address environmental challenges through regulation and policy.

4. Global Environmental Concerns (1970s-1980s):

- Concerns over air and water pollution, deforestation, and biodiversity loss gained global attention.
- The United Nations Conference on the Human Environment (Stockholm, 1972) marked a pivotal moment in international environmental diplomacy.

5. Conservation and Sustainable Development (1980s-1990s):

- The Brundtland Commission's report "Our Common Future" (1987) popularized the concept of sustainable development, emphasizing the need to balance economic growth with environmental protection.
- The Montreal Protocol (1987) addressed ozone depletion by phasing out the production of harmful substances.

6. Climate Change and Kyoto Protocol (1990s-2000s):

- Scientific consensus on anthropogenic climate change led to international efforts to address greenhouse gas emissions.
- The Kyoto Protocol (1997) aimed to reduce emissions from industrialized countries, though its effectiveness faced challenges.

7. Rise of NGOs and Grassroots Movements (Late 20th - Early 21st Century):

- Environmental organizations like Greenpeace, WWF, and Friends of the Earth gained prominence, advocating for policy change and conservation efforts.

- Grassroots movements and community-based initiatives played a pivotal role in addressing local environmental issues.

8. Global Agreements and Sustainable Development Goals (21st Century):

- The Paris Agreement (2015) brought countries together to address climate change by committing to emission reduction targets.
- The United Nations Sustainable Development Goals (SDGs) include goals related to environmental sustainability, poverty reduction, and social equity.

9. Emphasis on Environmental Justice and Intersectionality (21st Century):

- The environmental movement increasingly focuses on social justice, recognizing that environmental issues disproportionately affect marginalized communities.
- Intersectionality, recognizing the interconnectedness of various forms of oppression, has become a central theme in environmental advocacy.

The environmental movement's evolution showcases humanity's growing awareness of the interconnectedness of Earth's ecosystems and the need for collective action. From local conservation efforts to global agreements, the movement continues to evolve in response to new challenges and opportunities, shaping policies and practices that aim to secure a sustainable future for the planet and its inhabitants.

Profiles of key environmentalists and their contributions to global awareness and action

Numerous dedicated environmentalists have made significant contributions to raising awareness and driving action to protect the planet. Here are profiles of a few key figures and their notable contributions:

1. Rachel Carson (1907-1964): Rachel Carson was a marine biologist and writer whose groundbreaking book "Silent Spring" (1962) highlighted the dangers of pesticides, particularly DDT, to ecosystems and human health. Her work is often credited with sparking the modern environmental movement and influencing the ban of DDT and the creation of the U.S. Environmental Protection Agency.

2. Wangari Maathai (1940-2011): Wangari Maathai, a Kenyan environmental and political activist, founded the Green Belt Movement in 1977. This movement focused on tree planting, environmental conservation, and women's empowerment. Maathai's efforts led to the planting of millions of trees, soil conservation projects, and advocacy for women's rights and environmental justice.

3. Al Gore: Former U.S. Vice President Al Gore's documentary "An Inconvenient Truth" (2006) brought widespread attention to the issue of climate change. The film presented compelling scientific evidence of human-induced global warming and played a significant role in raising public awareness about the urgency of addressing climate change.

4. Greta Thunberg: Swedish climate activist Greta Thunberg gained international recognition for her school strike for climate action in 2018. Her "Fridays for Future" movement inspired millions of young people worldwide to participate in climate strikes and demand action from governments to address the climate crisis.

5. Jane Goodall: Primatologist and conservationist Jane Goodall is renowned for her groundbreaking research on chimpanzees and her efforts in wildlife conservation. The Jane Goodall Institute, founded by Goodall, focuses on chimpanzee protection, habitat conservation, and empowering local communities.

6. Vandana Shiva: Indian environmental activist Vandana Shiva is known for her advocacy against genetically modified crops, corporate control of agriculture, and promotion of sustainable farming practices. She founded Navdanya, an organization promoting seed saving, biodiversity, and sustainable agriculture.

7. Sir David Attenborough: Renowned British natural historian and broadcaster Sir David Attenborough has brought the wonders of the natural world into homes across the globe through his documentaries. His work, such as "Blue Planet II" and "Planet Earth," has raised awareness about conservation issues, ocean pollution, and the need to protect Earth's biodiversity.

8. Leonardo DiCaprio: Actor and environmentalist Leonardo DiCaprio established the Leonardo DiCaprio Foundation to support environmental causes. He has been a vocal advocate for addressing climate change, promoting renewable energy, and protecting biodiversity.

These profiles highlight only a few of the many individuals who have dedicated their lives to raising awareness and advocating for the protection of the environment. Their contributions have inspired global action, policy changes, and a greater understanding of the urgent need to safeguard the planet for current and future generations.

Conservation and its importance in maintaining balance in ecosystems

Conservation plays a vital role in maintaining the delicate balance within ecosystems and ensuring the long-term health and sustainability of the natural world. Here's why conservation is so important:

1. Biodiversity Preservation: Conservation efforts aim to protect the rich variety of species within ecosystems. Biodiversity is essential for ecosystem stability, resilience, and adaptability. Each species has a specific role (niche) within the ecosystem, and the loss of even one species can disrupt the entire balance, potentially leading to a cascade of negative impacts.

2. Ecosystem Services: Ecosystems provide invaluable services that directly benefit human well-being. These services include clean air and water, pollination of crops, regulation of climate, nutrient cycling, and disease control. Conserving ecosystems ensures the continuation of these services, which are vital for agriculture, health, and economic activities.

3. Habitat Protection: Conservation efforts focus on protecting and restoring natural habitats. Habitats provide essential resources for plants and animals, including food, shelter, and breeding grounds. When habitats are destroyed or degraded, species are displaced, and the intricate relationships between organisms are disrupted.

4. Genetic Diversity: Genetic diversity within species is crucial for their ability to adapt to changing conditions, such as climate shifts or disease outbreaks. Preserving genetic diversity enhances the resilience of populations and helps prevent inbreeding and the associated genetic problems.

5. Ecosystem Resilience: Conserved ecosystems are better equipped to withstand disturbances like natural disasters, disease outbreaks, and climate change. High biodiversity and intact habitats provide a buffer against sudden changes, allowing ecosystems to recover more quickly.

6. Cultural and Aesthetic Value: Many cultures hold deep connections to their local environments, and natural landscapes have cultural and aesthetic significance. Conservation ensures that future generations can appreciate and benefit from the beauty and cultural heritage of these areas.

7. Scientific Research and Discovery: Conserved areas provide valuable opportunities for scientific research and discovery. Studying pristine ecosystems helps us understand ecological processes, species interactions, and the impacts of human activities on the environment. This knowledge informs conservation strategies and management practices.

8. Climate Change Mitigation: Healthy ecosystems, such as forests and wetlands, act as carbon sinks, absorbing and storing carbon dioxide from the atmosphere. Conserving these areas helps mitigate the impacts of climate change by reducing greenhouse gas emissions and preserving natural carbon storage.

9. Prevention of Extinctions: Conservation efforts can prevent the extinction of endangered and threatened species. Each species has its own unique contribution to the ecosystem, and losing a species can disrupt ecological functions and result in cascading effects.

10. Ethical Responsibility: Humans share the planet with countless other species, and there's an ethical responsibility to ensure their survival and well-being. Conservation reflects our stewardship of the Earth and demonstrates a commitment to the ethical treatment of all living beings.

In essence, conservation is essential for maintaining the intricate web of life that sustains ecosystems and supports human societies. By protecting biodiversity, preserving habitats, and promoting sustainable practices, we contribute to the resilience and health of Earth's ecosystems for present and future generations.

Case studies of successful conservation efforts and the positive impacts they've had

Several successful conservation efforts have demonstrated the positive impact of dedicated actions and policies. These case studies showcase the effectiveness of conservation in preserving biodiversity, restoring ecosystems, and benefiting both the environment and local communities:

1. The Recovery of the California Condor: The California condor, one of the world's most endangered birds, faced extinction due to habitat loss, lead poisoning, and other threats. In the 1980s, a decision was made to capture all remaining wild condors for a breeding program. After decades of intensive management, captive breeding, and habitat protection, the population gradually increased. Today, California condors are still critically endangered but are slowly being reintroduced into the wild, demonstrating the potential for recovery even in the face of extreme challenges.

2. The Reforestation of Loess Plateau, China: The Loess Plateau in China had suffered from severe soil erosion and deforestation, leading to reduced agricultural productivity and devastating floods. In response, the Chinese government initiated the "Grain for Green" program in the late 1990s. This program involved paying farmers to plant trees and convert marginal croplands into forests and grasslands. As a result, the area's ecosystem has been restored, soil erosion has decreased, and downstream flooding has been significantly reduced.

3. The Protection of Gorongosa National Park, Mozambique: Gorongosa National Park in Mozambique faced severe challenges due to civil unrest and poaching. However, a collaborative effort involving the government, local communities, and international conservation organizations has resulted in remarkable progress. Poaching has been curbed, wildlife populations have rebounded, and efforts to engage local communities in conservation have led to improved livelihoods and reduced human-wildlife conflicts.

4. The Restoration of the Florida Everglades: The Florida Everglades, a unique wetland ecosystem, had been impacted by drainage for agriculture and urban development, leading to habitat loss and altered water flows. Restoration efforts, such as the Comprehensive Everglades Restoration Plan, focus on restoring natural water flows, reestablishing native vegetation, and improving water quality. These efforts aim to recover the ecosystem's health, benefit wildlife, and ensure a sustainable water supply for human populations.

5. The Protection of the Galápagos Islands, Ecuador: The Galápagos Islands faced threats from invasive species, tourism impacts, and overfishing. The Ecuadorian government and conservation organizations implemented strict regulations to control tourism, eradicate invasive species, and establish marine protected areas. These efforts have helped protect the unique and fragile ecosystems of the Galápagos, ensuring the survival of iconic species like the giant tortoise and blue-footed booby.

These case studies highlight the positive outcomes of conservation efforts driven by collaboration between governments, communities, scientists, and organizations. They demonstrate that with the right strategies, dedication, and long-term commitment, it is possible to reverse environmental degradation, restore ecosystems, and safeguard biodiversity for future generations.

Ecological principles and the relationships between organisms and their environments

Ecological principles encompass fundamental concepts that explain the interactions between organisms and their environments. These principles provide insights into the dynamics of ecosystems, species relationships, and the intricate balance that sustains life on Earth. Here are some key ecological principles and the relationships they describe:

1. **Interdependence and Interconnectedness:** All organisms within an ecosystem are interconnected and interdependent. Each species relies on others for resources, such as food, shelter, and breeding sites. Changes in one species can trigger a chain reaction that affects the entire ecosystem.

2. **Biotic and Abiotic Factors:** Ecosystems consist of both living (biotic) and non-living (abiotic) components. Biotic factors include all living organisms, while abiotic factors encompass physical elements like sunlight, water, temperature, soil, and nutrients. The interactions between these factors shape the structure and functioning of ecosystems.

3. **Habitat and Niche:** A habitat is the physical environment where a species lives, including all the biotic and abiotic factors. A niche refers to the role and position a species occupies within its habitat. It involves the species' interactions with other species, its behaviors, and its resource use.

4. **Trophic Levels and Energy Flow:** Organisms in an ecosystem are organized into trophic levels based on their energy source and

role in the food chain. Producers (plants) capture energy from the sun through photosynthesis and form the base of the food chain. Consumers (herbivores, carnivores, omnivores) feed on producers or other consumers, and decomposers break down organic matter to recycle nutrients.

5. Food Chains and Food Webs: Food chains illustrate the flow of energy and nutrients from one organism to another. Food webs are interconnected chains that reflect the complex relationships within an ecosystem, as most organisms have multiple interactions and roles.

6. Competition and Resource Partitioning: Species within an ecosystem often compete for limited resources such as food, territory, and mates. Resource partitioning occurs when species adapt to minimize competition by using different resources or occupying different niches within the same habitat.

7. Predation and Herbivory: Predation involves one species (predator) feeding on another species (prey). This relationship can regulate prey populations and influence community dynamics. Similarly, herbivores consume plant material, shaping plant population sizes and distribution.

8. Symbiotic Relationships: Symbiotic relationships involve interactions between different species that live in close proximity. Mutualism benefits both species, commensalism benefits one without harming the other, and parasitism benefits one while harming the other.

9. Succession and Disturbance: Ecological succession is the gradual change in species composition and community structure over time in response to changing environmental conditions. Disturbances, like fires or storms, can reset succession and create opportunities for new species to establish themselves.

10. Ecosystem Services: Ecosystems provide essential services, such as air and water purification, pollination, nutrient cycling, and climate regulation, which are vital for human well-being.

These ecological principles underscore the complexity and beauty of the relationships that shape ecosystems. Understanding these principles is essential for effective conservation, management, and sustainable use of natural resources.

Food chains, food webs, and the role of keystone species

Food Chains and Food Webs: Food chains and food webs describe the flow of energy and nutrients through ecosystems.

- **Food Chains:** A food chain is a linear sequence that shows how energy and nutrients are transferred from one organism to another through consumption. It typically starts with a producer (plants), followed by primary consumers (herbivores), secondary consumers (carnivores), and sometimes tertiary consumers (top predators).
- **Food Webs:** A food web is a more complex representation of feeding relationships within an ecosystem. It includes multiple interconnected food chains and shows the interdependence of various species. Organisms often occupy multiple trophic levels, and their interactions are depicted as arrows pointing from prey to predator.

Role of Keystone Species: A keystone species is a species that has a disproportionately large impact on its ecosystem relative to its abundance or biomass. Removing or altering a keystone species can cause dramatic changes in the structure and functioning of an entire ecosystem. Here's why keystone species are crucial:

1. **Biodiversity Regulation:** Keystone species help maintain biodiversity by preventing any single species from dominating the ecosystem. They influence the distribution and abundance of other species, promoting

a balanced community.

2. **Ecosystem Structure:** Keystone species can shape the physical structure of an ecosystem. For instance, beavers are considered keystone species because their dam-building activities create wetland habitats that benefit a wide range of organisms.
3. **Predator Regulation:** Some keystone species are predators that control the populations of their prey, preventing them from becoming too abundant and outcompeting other species. Sea otters in kelp forest ecosystems are an example of this.
4. **Resource Availability:** Keystone species can indirectly influence other species by affecting resource availability. For instance, a predator that feeds on a dominant competitor can create space for other species to thrive.
5. **Trophic Cascade:** Keystone species can trigger a "trophic cascade," where changes in one trophic level influence multiple other levels. For example, the removal of top predators can lead to an overabundance of herbivores, which in turn affects plant populations.
6. **Ecosystem Stability:** The presence of keystone species enhances ecosystem stability by preventing excessive fluctuations in species abundance and maintaining ecological balance.

Example of a Keystone Species: Sea Otters: Sea otters are considered a keystone species in kelp forest ecosystems. They feed on sea urchins, which are herbivores that graze on kelp. Without sea otters, sea urchin populations can explode, leading to overgrazing of kelp forests. This impacts the entire ecosystem, as kelp forests provide habitat and food for numerous species. The presence of sea otters helps maintain the health and balance of kelp forest ecosystems.

Keystone species exemplify the interconnectedness of ecosystems and the profound influence that individual species can have on the structure and functioning of the entire community.

Significance of recycling in reducing waste and conserving resources

Recycling plays a crucial role in reducing waste, conserving valuable resources, and promoting sustainable practices. Its significance extends beyond waste management, as recycling contributes to environmental protection, energy savings, and the reduction of greenhouse gas emissions. Here are some key reasons why recycling is important:

1. Resource Conservation: Recycling helps conserve natural resources such as minerals, metals, timber, and fossil fuels. By reusing materials, we reduce the need for extracting and processing virgin resources, which often involves energy-intensive and environmentally damaging processes.

2. Energy Savings: Recycling typically requires less energy compared to producing goods from raw materials. For instance, recycling aluminum uses significantly less energy than producing new aluminum from bauxite ore. Energy savings translate to reduced greenhouse gas emissions and contribute to efforts to combat climate change.

3. Waste Reduction: Recycling diverts materials from landfills and incineration, reducing the amount of waste that ends up polluting the environment. This helps extend the lifespan of landfills and decreases the release of harmful pollutants and toxins.

4. Lower Pollution: Extracting and processing raw materials can lead to air and water pollution, habitat destruction, and soil degradation. Recycling reduces the need for these processes, which helps protect ecosystems and minimize environmental damage.

5. Reduced Greenhouse Gas Emissions: Manufacturing goods from recycled materials generates fewer greenhouse gas emissions compared to using virgin resources. Recycling reduces the carbon footprint associated with resource extraction, transportation, and production.

6. Preservation of Ecosystems: Reducing the demand for raw materials helps protect ecosystems and biodiversity. For example, recycling paper reduces the need to cut down forests, which are essential habitats for many species.

7. Economic Benefits: Recycling industries create jobs and contribute to the economy. Recycling also decreases the costs associated with waste disposal, as recycling programs can be more cost-effective than managing landfills and incineration facilities.

8. Promoting Circular Economy: Recycling is a key component of the circular economy, where products and materials are designed, produced, and used in a way that minimizes waste and encourages reuse and recycling. This approach reduces the "take, make, dispose" linear model of production and consumption.

9. Consumer Awareness and Education: Recycling raises awareness about the impact of consumption on the environment and encourages responsible consumer behavior. It encourages people to consider the lifecycle of products and make choices that align with sustainability.

10. Long-Term Sustainability: Recycling is a practical step towards achieving a sustainable future. It helps ensure that resources are available for current and future generations and promotes responsible stewardship of the planet.

In essence, recycling is a cornerstone of environmental stewardship and a tangible way for individuals, communities, and industries to contribute to a more sustainable and resilient future.

Innovative recycling programs and the benefits of adopting sustainable practices

Innovative recycling programs and sustainable practices are playing a crucial role in addressing waste management challenges, conserving resources, and promoting environmental well-being. These programs demonstrate the potential of creative solutions to contribute to a more sustainable future. Here are some examples of innovative recycling programs and the benefits they offer:

1. E-Waste Recycling: E-waste recycling programs focus on recovering valuable materials from electronic devices such as smartphones, laptops, and appliances. These programs prevent hazardous materials from entering landfills and incinerators while recovering precious metals like gold and rare earth elements. Additionally, e-waste recycling reduces the demand for virgin resources used in electronic manufacturing.

2. Plastic Bottle Recycling and Upcycling: Plastic bottle recycling programs have led to the creation of sustainable products like recycled polyester clothing, backpacks, and even building materials. Some innovative companies are transforming plastic waste into fashionable items, reducing the reliance on fossil fuels for new plastic production and diverting plastic from the waste stream.

3. Composting and Organic Waste Management: Community composting programs enable residents to divert organic waste from landfills and turn it into nutrient-rich compost. This compost can then be used to enrich soil for gardening and agriculture. Organic waste diversion reduces methane emissions from landfills and supports healthier soils and plant growth.

4. Food Rescue Programs: Food rescue initiatives collect surplus and unsold edible food from restaurants, grocery stores, and events to redistribute it to people in need. These programs combat food waste, reduce landfill contributions, and address hunger and food insecurity.

5. Clothing and Textile Recycling: Textile recycling programs encourage people to donate used clothing and textiles. Some organizations recycle these materials into new clothing or repurpose them for various applications, reducing the environmental impact of fast fashion and the waste generated by discarded textiles.

6. Precious Metal Recovery from Electronics: Innovative methods are being developed to recover precious metals like gold, silver, and palladium from discarded electronic devices. These processes help reduce the need for mining and extraction, which can have significant environmental and social impacts.

Benefits of Adopting Sustainable Practices:

1. **Environmental Preservation:** Sustainable practices reduce the consumption of resources, minimize pollution, and protect ecosystems and biodiversity.
2. **Waste Reduction:** Adopting sustainable practices reduces waste generation, which lessens the burden on landfills and incinerators.
3. **Energy and Resource Conservation:** Sustainable practices often involve using fewer resources and less energy, leading to lower carbon emissions and a smaller ecological footprint.
4. **Innovation and Economic Growth:** The transition to sustainable practices drives innovation, creates new markets for eco-friendly products, and generates jobs in sectors like renewable energy and green technology.

5. **Public Health Improvement:** Reducing pollution, waste, and harmful chemicals benefits public health by decreasing air and water pollution and minimizing exposure to toxins.

6. **Resilience to Climate Change:** Sustainable practices contribute to building more resilient communities and ecosystems that can better withstand the impacts of climate change.

7. **Social Responsibility:** Adopting sustainable practices demonstrates a commitment to ethical and responsible behavior, appealing to environmentally conscious consumers and stakeholders.

8. **Long-Term Viability:** Sustainable practices help ensure the long-term viability of resources and ecosystems for future generations.

By implementing innovative recycling programs and adopting sustainable practices, individuals, businesses, and communities can contribute to a healthier planet, a more vibrant economy, and a better quality of life for all.

Importance of clean water sources for both human and ecological health

Clean water sources are essential for both human and ecological health. Access to clean water is a fundamental human right and a critical component of maintaining thriving ecosystems. Here's why clean water is so important:

Importance for Human Health:

1. **Drinking Water:** Clean and safe drinking water is essential for human survival and well-being. Contaminated water can lead to waterborne diseases such as cholera, dysentery, and typhoid, which pose significant health risks, particularly in areas with inadequate sanitation and water treatment.
2. **Hygiene and Sanitation:** Clean water is crucial for personal hygiene, sanitation, and preventing the spread of infectious diseases. Adequate water supply ensures proper handwashing, hygiene practices, and sanitation facilities, which are essential for disease prevention.
3. **Nutrition and Agriculture:** Clean water is essential for agriculture and food production. It is used for irrigation, livestock watering, and food processing. Access to clean water improves crop yield and contributes to food security.
4. **Economic Development:** Reliable access to clean water is crucial for economic development. Industries, businesses, and tourism rely on clean water sources for production, operations, and attracting visitors.

5. **Reduced Healthcare Burden:** Ensuring clean water sources can reduce the burden on healthcare systems by preventing waterborne diseases and reducing the need for medical treatment.
6. **Children's Health:** Clean water is especially critical for the health of children, who are more susceptible to waterborne illnesses. Access to clean water in schools supports a healthy learning environment.

Importance for Ecological Health:

1. **Aquatic Ecosystems:** Aquatic ecosystems, such as rivers, lakes, and oceans, rely on clean water for maintaining biodiversity, habitat quality, and ecosystem stability. Polluted water can harm aquatic life, disrupt food chains, and lead to declines in fish populations.
2. **Water Quality:** Clean water is necessary for maintaining water quality in aquatic systems. Polluted water can lead to oxygen depletion, algal blooms, and contamination of aquatic habitats.
3. **Wetlands and Riparian Zones:** Wetlands and riparian zones play a critical role in filtering and purifying water. They act as natural buffers, removing pollutants and sediment before they reach water bodies.
4. **Wildlife Habitat:** Clean water sources provide habitats for diverse species of plants and animals. Wetlands, rivers, and lakes are essential breeding and feeding grounds for many species.
5. **Ecosystem Services:** Healthy aquatic ecosystems provide essential ecosystem services such as water purification, flood regulation, and climate regulation, which benefit both humans and nature.
6. **Economic Value:** Clean water sources support recreational activities like fishing, boating, and swimming, contributing to local economies and

tourism industries.

7. **Biodiversity Conservation:** Protecting clean water sources is essential for conserving biodiversity and preserving the natural balance of ecosystems.

In essence, clean water is a fundamental requirement for life and is crucial for human health, economic development, and the well-being of the natural world. Ensuring the availability and quality of clean water sources is a shared responsibility that requires sustainable management, conservation, and cooperation among governments, communities, and industries.

Water scarcity, pollution, and the global efforts to ensure safe water access

Water scarcity and pollution are significant challenges that affect communities around the world. Addressing these issues requires global efforts to ensure safe water access, protect water resources, and promote sustainable water management practices. Here's an overview of the challenges and global initiatives in place:

Water Scarcity: Water scarcity refers to the imbalance between water availability and the demand for water. It can result from factors such as population growth, climate change, inefficient water use, and inadequate water management. Water scarcity has far-reaching implications for human well-being, agriculture, industry, and ecosystems.

Water Pollution: Water pollution occurs when contaminants, such as chemicals, pathogens, and waste, are introduced into water bodies, making the water unsafe for human consumption and harming aquatic ecosystems. Polluted water can lead to health issues, ecosystem degradation, and the loss of biodiversity.

Global Efforts to Ensure Safe Water Access:

1. **United Nations Sustainable Development Goal 6 (SDG 6):** SDG 6 aims to ensure access to clean water and sanitation for all by 2030. It emphasizes the importance of water quality, efficiency, and equitable access. Efforts are focused on improving water infrastructure, sanitation facilities, and sustainable water management.

2. **The Water Action Decade (2018-2028):** The United Nations launched the Water Action Decade to accelerate efforts towards achieving water-related SDGs. It encourages collaboration among governments, organizations, and communities to address water challenges.

3. **International Water Treaties and Agreements:** Various international agreements aim to promote sustainable water use and transboundary water management. Examples include the United Nations Watercourses Convention and the Ramsar Convention on Wetlands.

4. **World Water Day:** Observed annually on March 22nd, World Water Day raises awareness about water-related issues and promotes sustainable water management practices.

5. **Water Conservation and Efficiency:** Many countries are implementing water conservation measures to reduce water waste and improve efficiency in agriculture, industry, and households.

6. **Investment in Water Infrastructure:** Governments and organizations are investing in water infrastructure projects, such as building dams, reservoirs, and wastewater treatment plants, to improve water availability and quality.

7. **Educational and Awareness Campaigns:** Numerous initiatives focus on educating communities about water conservation, pollution prevention, and the importance of safe water access.

8. **Integrated Water Resource Management (IWRM):** IWRM is a holistic approach that considers social, economic, and environmental factors in water management. It emphasizes stakeholder involvement, sustainable use, and equitable access.

9. **Research and Innovation:** Advances in water treatment technologies, desalination, and water recycling are helping address water scarcity and pollution challenges.

10. **Collaborative Partnerships:** Governments, non-governmental organizations, businesses, and communities are working together to develop and implement solutions for water-related challenges.

Global efforts to ensure safe water access require cooperation among nations, strong governance, effective policies, technological innovation, and community engagement. By addressing water scarcity and pollution, the world can safeguard the health and well-being of current and future generations, promote sustainable development, and protect the integrity of Earth's ecosystems.

Earth's climate system, atmospheric phenomena, and weather patterns

Earth's climate system is a complex and interconnected network of physical, chemical, and biological processes that regulate the planet's temperature, weather patterns, and overall climate. Atmospheric phenomena and weather patterns are integral components of this system, shaping Earth's climate on various scales. Here's an overview of these concepts:

Earth's Climate System: The climate system consists of several interacting components, including the atmosphere, hydrosphere (oceans, lakes, rivers), cryosphere (ice and snow), biosphere (living organisms), and lithosphere (Earth's solid outer layer). These components exchange energy, moisture, and materials, leading to dynamic interactions that influence climate patterns and trends.

Atmospheric Phenomena: The atmosphere is the layer of gases surrounding Earth. Various atmospheric phenomena play a crucial role in shaping Earth's climate:

1. **Greenhouse Effect:** The greenhouse effect is the process by which certain gases in the atmosphere (such as carbon dioxide and water vapor) trap heat from the Sun, keeping Earth's temperature within a range suitable for life. However, human activities, such as burning fossil fuels, have increased the concentration of these greenhouse gases, leading to enhanced global warming.

2. **Solar Radiation:** The Sun emits solar radiation, which heats Earth's surface. The amount of solar radiation received by different regions varies based on factors like

latitude and time of year, contributing to temperature differences and atmospheric circulation.

3. **Atmospheric Circulation:** The movement of air due to differences in temperature and pressure drives atmospheric circulation patterns. This includes the Hadley, Ferrel, and Polar cells, as well as the jet streams. These patterns influence weather systems and play a role in distributing heat around the globe.

4. **Trade Winds and Monsoons:** Trade winds are prevailing winds that blow from east to west in the tropics. Monsoons are seasonal winds that bring heavy rainfall. These patterns significantly affect weather and climate in regions like Southeast Asia and Africa.

Weather Patterns: Weather refers to short-term atmospheric conditions in a specific location, including temperature, humidity, wind, and precipitation. Weather patterns are influenced by a variety of factors:

1. **Air Masses and Fronts:** Air masses are large bodies of air with consistent temperature and moisture characteristics. Fronts are boundaries between different air masses. Interactions between air masses and fronts lead to changes in weather conditions.

2. **Cyclones and Anticyclones:** Cyclones are low-pressure systems that bring clouds and precipitation, while anticyclones are high-pressure systems associated with clear skies and fair weather.

3. **Clouds and Precipitation:** Different types of clouds (cumulus, stratus, cirrus) and precipitation (rain, snow, sleet, hail) result from varying atmospheric conditions, temperature, and moisture levels.

4. **El Niño and La Niña:** These climate phenomena, known as the El Niño-Southern Oscillation (ENSO), are characterized by changes in sea surface temperatures in the Pacific Ocean. They have far-reaching effects on

global weather patterns, leading to droughts, floods, and temperature anomalies in various regions.

Understanding Earth's climate system, atmospheric phenomena, and weather patterns is crucial for predicting climate change, managing natural resources, and adapting to changing environmental conditions. Ongoing research and monitoring efforts help us comprehend these complex interactions and their implications for the planet's future.

Impact of climate change on weather events and the planet as a whole

Climate change has significant and far-reaching impacts on weather events and the entire planet. As the Earth's average temperature rises due to human activities, such as the burning of fossil fuels and deforestation, it disrupts the planet's natural climate patterns and leads to a range of consequences. Here's how climate change affects weather events and the planet as a whole:

1. Extreme Weather Events: Climate change increases the frequency and intensity of various extreme weather events:

- **Heatwaves:** Rising temperatures contribute to more frequent and severe heatwaves, which can have deadly health impacts, strain energy systems, and damage crops.
- **Heavy Rainfall and Flooding:** Warmer air holds more moisture, leading to more intense rainfall events and an increased risk of flooding in many regions.
- **Droughts:** Changes in precipitation patterns and increased evaporation can lead to prolonged and more severe droughts, affecting agriculture, water supply, and ecosystems.
- **Hurricanes and Cyclones:** While the relationship between climate change and hurricane intensity is complex, warming oceans can lead to stronger and potentially more destructive hurricanes and cyclones.

2. Sea Level Rise: Melting glaciers and thermal expansion of seawater due to warming oceans contribute to rising sea levels.

This leads to:

- **Coastal Erosion:** Higher sea levels intensify coastal erosion and increase the risk of property damage and loss of infrastructure.
- **Increased Flooding:** Rising sea levels exacerbate the impacts of storm surges, causing more frequent and severe coastal flooding during storms.

3. Glacier and Ice Melt: Warming temperatures lead to the melting of glaciers and ice sheets in polar regions and mountain ranges:

- **Sea Ice Loss:** Declining Arctic sea ice affects ecosystems and contributes to changes in weather patterns and ocean circulation.
- **Glacial Retreat:** Glacial melt contributes to rising sea levels and alters freshwater availability for communities dependent on glacier-fed rivers.

4. Ocean Changes: Climate change affects ocean currents, temperatures, and chemistry:

- **Ocean Warming:** Warmer oceans can disrupt marine ecosystems, affecting marine life distribution, migration patterns, and coral reefs.
- **Ocean Acidification:** Increased carbon dioxide levels lead to higher acidity in oceans, which negatively impacts marine life that relies on calcium carbonate for shells and skeletons.

5. Ecosystem Disruption: Climate change disrupts ecosystems and biodiversity:

- **Species Migration:** Species may shift their ranges to find suitable conditions, affecting the balance of ecosystems.
- **Coral Bleaching:** Warmer ocean temperatures can cause coral bleaching, leading to the degradation of coral reefs and the ecosystems they support.

6. Agriculture and Food Security: Changing weather patterns can impact crop yields and food supply:

- **Crop Failures:** Increased heat, droughts, and extreme weather events can lead to reduced agricultural productivity, affecting food availability and prices.
- **Crop Disease:** Warmer temperatures can promote the spread of pests and diseases that affect crops and livestock.

Addressing the impacts of climate change requires global efforts to reduce greenhouse gas emissions, transition to renewable energy sources, and implement sustainable land and water management practices. Adaptation strategies are also crucial to minimize the impacts on communities and ecosystems that are already facing the consequences of a changing climate.

Causes and consequences of desertification

Causes of Desertification: Desertification is the process by which fertile land becomes desert due to various factors, including both natural and human-induced processes. Some of the key causes of desertification include:

1. **Climate Change:** Changes in climate patterns, including increased temperatures and altered precipitation patterns, can lead to reduced soil moisture and increased evaporation, contributing to desertification.

2. **Deforestation:** Clearing forests for agriculture, grazing, and fuelwood reduces vegetation cover, leading to soil erosion, decreased soil fertility, and increased vulnerability to desertification.

3. **Overgrazing:** Excessive grazing by livestock can strip vegetation cover, compact soil, and lead to soil degradation, making the land more susceptible to erosion and desertification.

4. **Unsustainable Agricultural Practices:** Poor land management, improper irrigation techniques, and excessive use of agrochemicals can degrade soil quality and contribute to desertification.

5. **Soil Erosion:** Wind and water erosion can remove the fertile topsoil layer, leaving behind less productive and less resilient soil.

6. **Water Mismanagement:** Over-extraction of groundwater and inefficient water use can deplete water resources and contribute to soil salinization, making the land less suitable for agriculture.

Consequences of Desertification: The consequences of

desertification are widespread and impact both the environment and human societies:

1. **Loss of Biodiversity:** Desertification reduces habitat suitability for many plant and animal species, leading to decreased biodiversity and the potential extinction of certain species.
2. **Food Insecurity:** Reduced agricultural productivity due to desertification can lead to food shortages, malnutrition, and increased dependence on food imports.
3. **Water Scarcity:** Desertification often leads to reduced water availability due to the degradation of water sources and decreased soil moisture retention.
4. **Economic Decline:** Desertification can lead to reduced incomes for farmers and pastoralists, loss of livelihoods, and increased poverty in affected regions.
5. **Migration and Displacement:** As land becomes less productive, people may be forced to migrate in search of better opportunities, which can lead to social and political tensions.
6. **Increased Vulnerability to Natural Disasters:** Degraded landscapes are more prone to floods, landslides, and other natural disasters, which can exacerbate the impacts of desertification.
7. **Loss of Ecosystem Services:** Desertification reduces the ability of ecosystems to provide services such as carbon sequestration, water purification, and erosion control.
8. **Climate Feedbacks:** Desertification can contribute to climate change by releasing stored carbon in soil and vegetation, further exacerbating global warming.

Efforts to combat desertification involve sustainable land management practices, reforestation, water conservation, soil conservation, and community engagement. International agreements, such as the United Nations Convention to Combat

Desertification (UNCCD), aim to address desertification and promote sustainable development in affected regions. Preventing and reversing desertification is crucial for maintaining ecosystem health, supporting local communities, and ensuring the long-term productivity of land resources.

Initiatives to combat desertification and restore degraded lands

Numerous initiatives and strategies have been developed to combat desertification and restore degraded lands. These efforts focus on sustainable land management, reforestation, soil conservation, water management, and community engagement. Here are some key initiatives and approaches:

1. **United Nations Convention to Combat Desertification (UNCCD):** The UNCCD is a global agreement aimed at combating desertification, land degradation, and drought. It promotes sustainable land management practices, emphasizes the importance of local participation, and encourages countries to develop action plans to address land degradation.

2. **Sustainable Land Management (SLM):** SLM involves adopting practices that improve land productivity while maintaining or enhancing the health of the land. These practices include agroforestry, crop rotation, contour farming, terracing, and no-till agriculture.

3. **Reforestation and Afforestation:** Planting trees and restoring forests in degraded areas can help improve soil structure, prevent erosion, increase water retention, and enhance biodiversity. Afforestation involves establishing forests in areas that were not previously forested, while reforestation involves restoring deforested or degraded areas.

4. **Soil Conservation Techniques:** Methods such as contour bunding, cover cropping, mulching, and erosion control measures

help prevent soil erosion, improve soil fertility, and enhance water retention.

5. Water Management: Efficient water management practices, such as rainwater harvesting, drip irrigation, and proper watershed management, help conserve and optimize water resources for agricultural and ecosystem needs.

6. Sustainable Grazing and Pasture Management: Implementing rotational grazing, establishing rest periods for pastures, and adopting managed grazing systems can help prevent overgrazing and land degradation in pastoral areas.

7. Agroforestry: Agroforestry combines tree planting with agricultural crops or livestock. It helps improve soil quality, provides shade, protects crops, and enhances biodiversity.

8. Soil Rehabilitation and Restoration: Techniques such as biochar application, organic matter incorporation, and soil amendment can help restore soil health and fertility in degraded areas.

9. Community-Based Initiatives: Engaging local communities in the decision-making process and involving them in land restoration efforts increases the chances of success. These initiatives often incorporate traditional knowledge and practices.

10. International Funding and Collaboration: International organizations, governments, and NGOs provide funding, technical support, and capacity-building for land restoration projects. Collaborative efforts facilitate knowledge exchange and sharing of best practices.

11. Research and Innovation: Research institutions and organizations work to develop and promote innovative technologies, methods, and approaches for land restoration and combating desertification.

12. Awareness and Education: Raising awareness about the importance of combating desertification and restoring degraded lands fosters public support and encourages behavioral changes.

By combining these initiatives and approaches, governments, organizations, communities, and individuals can work together to combat desertification, restore degraded lands, and ensure the sustainable use of land resources for current and future generations.

Occurrence and causes of natural disasters like earthquakes, hurricanes, and wildfires

Natural disasters, such as earthquakes, hurricanes, and wildfires, can have devastating impacts on communities, ecosystems, and infrastructure. These events are often driven by geological, meteorological, or environmental factors. Here's an overview of their occurrence and causes:

1. **Earthquakes: Occurrence:** Earthquakes occur due to the sudden release of energy along faults or fractures in the Earth's crust. These events can happen anywhere on Earth but are more common along tectonic plate boundaries. **Causes:** The primary cause of earthquakes is the movement of tectonic plates. This movement creates stress and strain along faults. When the accumulated stress exceeds the strength of the rocks, it's released as seismic energy, causing the ground to shake.

2. **Hurricanes (Tropical Cyclones or Typhoons): Occurrence:** Hurricanes, also known as tropical cyclones or typhoons in different regions, occur over warm ocean waters in tropical and subtropical regions. **Causes:** Hurricanes form when warm ocean waters evaporate and rise, creating low-pressure systems. As the warm, moist air rises, it cools and condenses, releasing heat and energy. This process causes the air to spiral around the low-pressure center, forming a hurricane.

3. **Wildfires: Occurrence:** Wildfires occur in forested, grassland, or shrubland areas where vegetation and fuel sources are abundant. They can be seasonal or occur year-round in regions prone to drought and dry conditions. **Causes:** Wildfires can be

ignited by both natural and human factors. Lightning strikes, volcanic activity, and spontaneous combustion can start natural wildfires. Human activities such as campfires, discarded cigarette butts, and power lines can also trigger fires.

It's important to note that while these natural disasters have natural triggers, human activities can exacerbate their impacts. For example, urbanization and development in earthquake-prone areas can increase vulnerability, deforestation can contribute to wildfire risks, and climate change can influence the intensity and frequency of hurricanes. Addressing the causes and minimizing the impacts of these disasters often involves a combination of preparedness, mitigation, and response strategies.

Disaster preparedness, mitigation strategies, and community resilience

Disaster preparedness, mitigation strategies, and community resilience are crucial components of reducing the impact of natural disasters and increasing the ability of communities to withstand and recover from these events. Here's an overview of each of these aspects:

1. **Disaster Preparedness:** Disaster preparedness involves planning, organizing, and equipping communities to effectively respond to and recover from disasters. It includes the following key elements:

- **Emergency Plans:** Developing comprehensive emergency plans that outline roles, responsibilities, and actions to take before, during, and after a disaster.
- **Communication:** Establishing communication networks and systems that enable timely dissemination of information and warnings to the public.
- **Evacuation Plans:** Creating evacuation routes, shelters, and protocols for safely moving residents away from disaster-prone areas.
- **Emergency Kits:** Encouraging individuals and families to have emergency supply kits containing essentials like water, food, medications, and first aid supplies.
- **Training and Drills:** Conducting regular training exercises and drills to ensure that individuals, first responders, and community members are familiar with emergency procedures.

2. Mitigation Strategies: Mitigation strategies focus on reducing the impact of disasters before they occur. These strategies aim to minimize damage to lives, property, and the environment:

- **Building Codes and Zoning:** Implementing and enforcing building codes and land-use zoning regulations that consider the potential risks posed by natural disasters.
- **Infrastructure Improvements:** Designing and constructing infrastructure to be more resilient against disaster impacts, such as earthquake-resistant buildings and flood-resistant drainage systems.
- **Vegetation Management:** Clearing and maintaining vegetation in wildfire-prone areas to reduce fuel loads and the risk of uncontrolled fires.
- **Ecosystem Restoration:** Restoring and maintaining natural ecosystems like wetlands and forests that provide natural buffers against disasters.
- **Early Warning Systems:** Developing and implementing early warning systems that provide timely information about approaching disasters, such as hurricanes, floods, and tsunamis.

3. Community Resilience: Community resilience refers to a community's ability to adapt, recover, and bounce back from disasters. It involves building a collective capacity to withstand and respond effectively:

- **Social Cohesion:** Strengthening community connections, communication, and cooperation to enhance collective problem-solving during disasters.
- **Education and Training:** Educating residents about disaster risks, preparedness, and response strategies to empower them to take proactive measures.
- **Local Leadership:** Empowering local leaders and organizations to take an active role in disaster planning,

response, and recovery efforts.

- **Cultural Preservation:** Integrating cultural practices and traditions into disaster management strategies to ensure they are culturally sensitive and effective.
- **Psychological Support:** Providing psychological support and resources to help individuals and communities cope with the emotional impacts of disasters.

Effective disaster management requires collaboration among governments, non-governmental organizations, communities, and individuals. By combining disaster preparedness, mitigation strategies, and community resilience-building, societies can minimize the impacts of disasters, protect lives and property, and promote a more resilient and sustainable future.

Importance of environmental responsibility and individual actions

Environmental responsibility and individual actions play a crucial role in addressing global environmental challenges and ensuring the sustainable future of our planet. Here are some key reasons why environmental responsibility and individual actions are of paramount importance:

1. **Conservation of Natural Resources:** Individual actions, such as reducing energy consumption, water usage, and waste generation, contribute to the conservation of finite natural resources. By using resources more efficiently, we help ensure their availability for future generations.

2. **Mitigation of Climate Change:** Collective individual efforts, such as reducing carbon emissions, using renewable energy sources, and adopting sustainable transportation options, contribute to mitigating climate change. These actions help slow down the pace of global warming and reduce its harmful impacts.

3. **Preservation of Biodiversity:** By avoiding activities that harm ecosystems and supporting conservation efforts, individuals can help protect the diversity of plant and animal species. Biodiversity is essential for ecosystem health, resilience, and the provision of ecosystem services.

4. **Reduction of Pollution and Waste:** Individuals can play a role in reducing pollution by using eco-friendly products, minimizing plastic usage, recycling, and properly disposing of hazardous materials. These actions contribute to cleaner air, water, and soil.

5. Support for Sustainable Practices: Consumer choices and behaviors influence markets and industries. By supporting sustainable products, businesses, and practices, individuals can drive positive changes in the way products are produced, packaged, and distributed.

6. Creation of Social Awareness and Norms: Individual actions and advocacy help raise awareness about environmental issues and create social norms that prioritize sustainability. This can lead to broader societal change and policy advancements.

7. Empowerment and Civic Engagement: Environmental responsibility empowers individuals to take an active role in their communities, encouraging them to engage with local government, advocate for policy changes, and contribute to environmental initiatives.

8. Impact on Future Generations: The choices we make today have a direct impact on the quality of life for future generations. By taking responsibility for our environmental impact, we contribute to a healthier and more sustainable world for our children and grandchildren.

9. Global Responsibility: Environmental challenges, such as climate change, deforestation, and pollution, are global in nature. Individual actions, when scaled up collectively, can contribute to global solutions and create a shared sense of responsibility for the planet.

10. Ethical and Moral Imperative: For many, caring for the environment is rooted in ethical and moral values. Individuals often feel a sense of duty to protect the Earth and its inhabitants for current and future generations.

In essence, environmental responsibility and individual actions are essential components of a holistic approach to addressing environmental issues. While systemic changes and policy shifts are crucial, the cumulative impact of millions of individual choices can create a significant positive ripple effect that contributes to a more sustainable and resilient world.

Embrace a harmonious relationship with nature and become stewards of the environment

Embracing a harmonious relationship with nature and becoming stewards of the environment is a transformative and essential mindset that recognizes the interconnectedness of all living beings and the planet. This perspective emphasizes our role as caretakers and guardians of the Earth's resources and ecosystems. Here's why embracing this mindset is crucial:

1. Interconnectedness: Every action we take, no matter how small, has ripple effects throughout the environment. By acknowledging our interconnectedness with nature, we become more mindful of the consequences of our choices.

2. Long-Term Sustainability: As stewards of the environment, we prioritize the health and well-being of the planet for current and future generations. This requires making decisions that support long-term sustainability rather than short-term gains.

3. Respect for Diversity: Nature is rich with diverse species and ecosystems. Embracing a harmonious relationship means valuing and protecting this biodiversity, recognizing that each species plays a unique role in maintaining the health of the planet.

4. Ethical Responsibility: A harmonious relationship with nature is rooted in ethical values that acknowledge the inherent worth of all life forms and the planet itself. As stewards, we respect and uphold these values in our actions and decisions.

5. Mitigation of Environmental Challenges: By becoming stewards, we actively participate in addressing environmental challenges such as climate change, pollution, and habitat loss. This proactive engagement is essential for creating positive change.

6. Mindful Consumption: As stewards, we consider the impact of our consumption patterns on the environment. This mindset encourages us to make conscious choices, reduce waste, and support sustainable practices.

7. Preservation of Cultural Heritage: Many cultures have strong connections to their natural surroundings. Embracing a harmonious relationship with nature honors and preserves cultural heritage and indigenous wisdom that is deeply intertwined with the environment.

8. Fostering Community Engagement: As stewards, we inspire and encourage others to take similar actions, creating a sense of shared responsibility and collective impact within communities.

9. Restoration and Regeneration: Stewards actively engage in the restoration and regeneration of ecosystems that have been degraded or damaged. This involves planting trees, restoring wetlands, and revitalizing habitats.

10. Connection to Well-Being: A harmonious relationship with nature has positive effects on our well-being. Spending time in nature, practicing conservation, and caring for the environment contribute to a sense of fulfillment and purpose.

Ultimately, embracing a harmonious relationship with nature and becoming stewards of the environment is a profound shift in mindset that transcends individual actions. It's a collective commitment to honoring and safeguarding the intricate web of life on Earth. By embodying this role, we can contribute to the preservation and flourishing of the planet for generations to come.

Ongoing efforts to protect and restore Earth's ecosystems

Efforts to protect and restore Earth's ecosystems are vital for maintaining biodiversity, mitigating climate change, and ensuring the health and well-being of both nature and humanity. These ongoing initiatives encompass a range of strategies and actions aimed at conserving and rejuvenating ecosystems. Here are some key examples:

1. **Protected Areas and Conservation Reserves:** Governments and organizations establish protected areas, such as national parks, wildlife reserves, and marine protected areas, to safeguard critical habitats and biodiversity. These areas offer refuge for species and provide opportunities for research, education, and ecotourism.

2. **Reforestation and Afforestation:** Efforts to combat deforestation involve replanting trees in degraded areas (reforestation) and establishing new forests in previously non-forested areas (afforestation). This helps sequester carbon, prevent soil erosion, and restore habitats.

3. **Habitat Restoration:** Restoring degraded ecosystems, such as wetlands, grasslands, and coral reefs, involves removing invasive species, replanting native vegetation, and addressing factors that caused degradation. This promotes biodiversity and ecosystem functionality.

4. **Sustainable Agriculture and Land Management:** Promoting sustainable farming practices, agroforestry, and organic farming helps conserve soil fertility, reduce chemical usage, and minimize

habitat destruction. Implementing proper land management techniques prevents soil erosion and degradation.

5. Marine Conservation and Restoration: Initiatives like marine protected areas, coral reef restoration, and sustainable fishing practices aim to protect marine ecosystems, combat overfishing, and reduce plastic pollution in oceans.

6. Indigenous and Local Knowledge: Incorporating indigenous and local knowledge in ecosystem management respects traditional practices and wisdom, often leading to more effective conservation strategies.

7. Climate-Smart Conservation: Efforts that combine conservation goals with climate change adaptation and mitigation strategies contribute to both ecosystem protection and reducing greenhouse gas emissions.

8. Sustainable Urban Planning: Implementing green infrastructure, creating urban parks, and prioritizing walkable cities contribute to healthier urban ecosystems and enhance residents' quality of life.

9. Education and Awareness: Raising public awareness about the importance of ecosystems, biodiversity, and sustainable practices fosters a sense of responsibility and encourages individual and collective action.

10. International Agreements and Collaboration: Global agreements such as the Convention on Biological Diversity (CBD) and the Paris Agreement on climate change facilitate international cooperation in protecting ecosystems and addressing global environmental challenges.

11. Wildlife Conservation and Anti-Poaching Efforts: Efforts to combat illegal wildlife trade, reduce poaching, and protect endangered species contribute to biodiversity conservation and ecosystem health.

12. Restoration of Wetlands and Watersheds: Rehabilitating wetlands and watersheds helps regulate water flow, improve water quality, and provide habitat for diverse species.

These efforts are interconnected and require collaboration among governments, NGOs, communities, businesses, and individuals. As ecosystems are intricate and delicate, a combination of strategies that address various aspects of protection and restoration is essential for creating a more sustainable and resilient planet.

Potential for positive change and collective action

The potential for positive change and collective action is immense when individuals, communities, organizations, and governments come together to address global challenges, protect the environment, and promote sustainability. Here are some reasons why collective action holds significant promise:

1. Amplified Impact: When individuals and groups join forces, their combined efforts have a far greater impact than individual actions alone. This synergy leads to larger-scale initiatives, broader reach, and more significant outcomes.

2. Shared Knowledge and Resources: Collective action enables the sharing of knowledge, expertise, and resources. Collaborative efforts leverage diverse perspectives and talents, resulting in innovative solutions and more effective strategies.

3. Advocacy and Policy Influence: Collective action has the power to influence policymakers, driving the creation and implementation of policies that support sustainability and environmental protection.

4. Behavior Change: Collective efforts can shape social norms and influence behavior change on a larger scale. When individuals see others taking positive actions, they are more likely to adopt similar practices.

5. Economic Impact: Collective actions can influence markets by creating demand for sustainable products and practices. This, in turn, encourages businesses to adopt more environmentally friendly approaches.

6. Global Movements: Collective action can spark global movements that draw attention to critical issues, leading to international awareness and cooperation.

7. Grassroots Initiatives: Community-led initiatives often have deep local roots and can foster a strong sense of ownership and commitment among participants.

8. Technology and Social Media: In the digital age, social media platforms and online networks enable rapid information dissemination, mobilizing people for collective action on a global scale.

9. Innovation and Collaboration: Collective action fosters collaboration between different stakeholders, encouraging cross-sectoral partnerships and innovative solutions.

10. Empowerment and Inclusivity: Collective action empowers individuals to contribute to positive change, giving them a sense of agency and ownership over the outcomes.

11. Long-Term Sustainability: Collective action supports the development of sustained, long-term initiatives that can address complex challenges over time.

12. Ripple Effects: The impact of collective action often extends beyond its immediate goals, creating ripple effects that inspire others to get involved and initiate their initiatives.

13. Adaptive Responses: Collective action enables communities to adapt to changing circumstances, making them more resilient in the face of challenges.

14. Cultural Shifts: When collective actions lead to cultural shifts in attitudes and behaviors, they contribute to lasting change in societal values.

By recognizing the potential of collective action and

fostering collaboration among various stakeholders, society can make significant strides toward achieving environmental sustainability, addressing climate change, and ensuring a healthier and more prosperous future for all.